Secondary School Algebra

Sujith Vijay

Dedicated to the loving memory of my brother Renjith Vijay (1985 - 2015)

Contents

Chapter 1

Arithmetic

1.1 Integer Arithmetic

The four fundamental operations of integer arithmetic are addition, subtraction, multiplication and division. Let us examine the properties of these operations in detail.

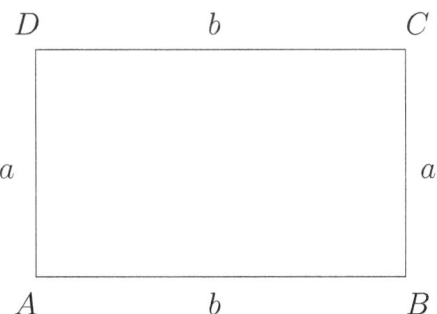

Let ABCD be a rectangle with sides a and b. The total distance travelled from A to C is the same whether we go via D or B. Thus $a + b = b + a$. We say that addition is **commutative**, in the sense that the answer does not depend on the order of the operands.

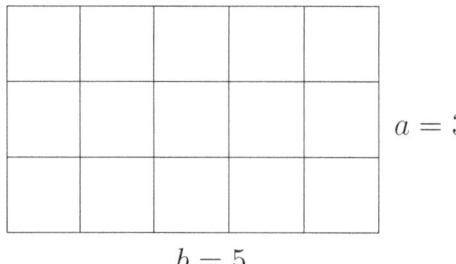

Consider a rectangular chessboard of a rows and b columns. The number of squares in the board is the same whether we count column-wise or row-wise. Thus $ab = ba$ and multiplication is also commutative.

However, subtraction and division are not commutative. For example, $1-2 \neq 2-1$ and $\frac{1}{2} \neq \frac{2}{1}$. In fact, $\frac{1}{2}$ is not even an element of the set of integers $\mathbb{Z} = \{0, \pm 1, \pm 2, \ldots\}$. Indeed, division is fundamentally different from addition, subtraction and multiplication in the context of integers. If a and b are integers, the sum $a + b$, the product ab and the difference $a - b$ are all integers. We say that **integers are closed under addition, subtraction and multiplication**. However, the ratio $\frac{1}{b}$ is an integer only if $b = 1$ or -1. Thus the set of integers is not closed under division.

The numbers 0 and 1 enjoy special status among all integers. These are known as the **additive identity** and the **multiplicative identity** respectively.

$$a + 0 = a = 0 + a \ \text{ for all } a \in \mathbb{Z}$$

$$a \cdot 1 = a = 1 \cdot a \ \text{ for all } a \in \mathbb{Z}$$

Moreover, 0 has a special property in the context of multiplication.

$$a \cdot 0 = 0 = 0 \cdot a \ \text{ for all } a \in \mathbb{Z}$$

The **additive inverse** of an integer a is the unique integer b with the property that $a + b = 0 = b + a$. For notational convenience, we write $b = -a$. We can now redefine subtraction as the process of adding the additive inverse.

$$a - b = a + (-b)$$

However, trying to define a multiplicative inverse in a similar manner, replacing the additive identity 0 with the multiplicative identity 1 will not always work. For example, there is no integer b with the property that $2 \cdot b = 1 = b \cdot 2$. The only integers with multiplicative inverses are 1 and -1 and each is its own multiplicative inverse. The essential difficulty is what we have already encountered, namely that integers are not closed under division.

Addition and multiplication are also **associative**. Thus we have,

$$(a + b) + c = a + (b + c)$$

$$(ab) \cdot c = a \cdot (bc)$$

What is the point of all this? Well, the sum of the digits of 1729 is 19. The product of the digits of 1729 is 126. But if you ask me the difference of the digits of

1729, I might offer to buy you a good breakfast. The combination of commutativity and associativity allows us to speak of *the sum* and *the product* of a set of integers in a way we cannot speak of *the difference*. In a commutative and associative world, all roads lead to Rome.

$$\begin{aligned} ((bd) \cdot a) \cdot c &= (a \cdot (bd)) \cdot c = ((ab) \cdot d) \cdot c = (ab) \cdot (dc) \\ &= (ab) \cdot (cd) = ((ab) \cdot c) \cdot d \end{aligned}$$

So far we have considered addition and multiplication as independent operations. The connection between the two is given by the **distributive property**.

$$a \cdot (b + c) = ab + ac$$

This can be extended to more than two operands as well. For example,

$$5 \cdot 3 = 5 \cdot (1 + 1 + 1) = 5 + 5 + 5$$

In general,

$$ab = a \cdot \underbrace{(1 + 1 + \cdots + 1)}_{b \text{ times}} = \underbrace{a + a + \cdots + a}_{b \text{ times}}$$

This is why we say that **multiplication is repeated addition**.

Multiplying an integer by -1 yields its additive inverse.

$$-a + a = 0 = 0 \cdot a = (-1 + 1) \cdot a = (-1) \cdot a + 1 \cdot a = (-1) \cdot a + a$$

4

Thus $(-1) \cdot a = -a$. (Here we are implicitly using the fact that if $x + z = y + z$, then it follows that $x = y$.)

It can also be shown that the product of two negative integers is a positive integer.

$$
\begin{aligned}
-(ab) + ab &= 0 = (-a + a) \cdot (-b + b) \\
&= (-a)(-b) - ab - ab + ab \\
&= -(ab) + (-a)(-b)
\end{aligned}
$$

Thus $(-a)(-b) = ab$. In particular, $(-1)(-1) = 1$. It follows that all even powers of -1 equal 1 and all odd powers of -1 equal -1.

The great joy in studying mathematics is that there is no need to take anything on blind faith. With the exception of the most fundamental axioms, every true statement can, at least in principle, be deduced by pure logical reasoning. There is no need to memorize anything.

1.2 Exponents and Powers

Just as multiplication is repeated addition, **exponentiation is repeated multiplication**. The notation a^m stands for the product $\underbrace{a \times a \times \cdots \times a}_{m \text{ times}}$. For example,

$$3^5 = 3 \times 3 \times 3 \times 3 \times 3 = 243.$$

Exponential notation using powers of 10 is very convenient to denote large as well as small numbers. For example,

$$10 \times 9 \times 8 \times \cdots \times 1 = 3628800 = 3.6288 \times 10^6$$

$$\frac{1}{64} = 0.015625 = 1.5625 \times 10^{-2}$$

It is clear from the definition that $a^m \times a^n = a^{m+n}$ and $(a^m)^n = a^{mn}$. Moreover, since multiplication is commutative, $(ab)^m = a^m \times b^m$.

These laws are sometimes helpful in comparing quantities. For example, can we determine which of the two numbers 19^{47} and 47^{19} is larger without computing their values? Yes, and here is how.

$$19^{47} > 16^{47} = (2^4)^{47} = 2^{188}$$

$$47^{19} < 64^{19} = (2^6)^{19} = 2^{114}$$

It follows that $19^{47} > 2^{188} > 2^{114} > 47^{19}$.

Since the identity $a^m = a^{m+0} = a^m \cdot a^0$ holds for all a and all m, a^0 must be the multiplicative identity, namely 1. One could argue that an exception must be made for $a = 0$, but professional mathematicians prefer to define $0^0 = 1$, in contrast to $\frac{0}{0}$ which is left as undefined. The motivation and advantages of such a definition are beyond the scope of this book. Suffice to say that $a^0 = 1$ for all a.

What about negative exponents? How should one define a^{-1}?

$a^{-1} \cdot a = a^{-1} \cdot a^1 = a^{-1+1} = a^0 = 1$. Therefore,

$$a^{-1} = \frac{1}{a} \text{ for all } a \neq 0$$

$$a^{-m} = (a^{-1})^m = \frac{1}{a^m} \text{ for all } a \neq 0$$

It now follows that

$$\frac{a^m}{a^n} = a^m \cdot \frac{1}{a^n} = a^m \cdot a^{-n} = a^{m-n} \text{ for all } a \neq 0$$

Moreover,
$$\frac{a^m}{b^m} = a^m \cdot \frac{1}{b^m} = a^m \cdot \left(\frac{1}{b}\right)^m = \left(\frac{a}{b}\right)^m \text{ for all } b \neq 0$$

Exponents can be fractions as well. For example, if a is any non-negative number, $a^{\frac{1}{2}}$ is the square root of a, since its square is given by

$$(a^{\frac{1}{2}})^2 = a^{\frac{2}{2}} = a^1 = a.$$

7

Recall that a **prime number** is a positive integer greater than 1 that cannot be expressed as the product of smaller integers. A positive integer greater than 1 is said to be a **composite number** if it is not a prime number. The **fundamental theorem of arithmetic** states that every composite number can be expressed as the product of powers of prime numbers and this representation is unique except for the order of factors. For example,

$$144 = 2^4 \cdot 3^2 \text{ and } 540 = 2^2 \cdot 3^3 \cdot 5^1$$

A positive integer is a perfect square if all the exponents appearing in the prime factorization are even, and a perfect cube if all exponents are multiples of 3. Thus 144 is a perfect square and 540 has to be multiplied by $3 \times 5 = 15$ to make it a perfect square. Similarly, 144 has to be multiplied by $2^2 \times 3 = 12$ to make it a perfect cube.

To estimate the square root or the cube root of an integer N, we find the two perfect squares closest to $4N$ or the two perfect cubes closest to $8N$. Thus if $N = 43$, then $8N = 344$ lies between $7^3 = 343$ and $8^3 = 512$. It follows that the cube root of 43 lies between 3.5 and 4, so the nearest integer to the cube root of 43 is 4. This may come as a surprise, as 43 is actually closer to $3^3 = 27$ than to $4^3 = 64$.

Decomposing composite numbers into their prime factors makes it easy to find the LCM and HCF of a pair of numbers. Suppose a and b are two positive integers, and the set of primes that appear in the prime factorization of at least one of them

8

is $\{p_1, p_2, \ldots, p_r\}$. We can then write

$$a = p_1^{e_1} p_2^{e_2} \cdots p_r^{e_r} \text{ and } b = p_1^{f_1} p_2^{f_2} \cdots p_r^{f_r}$$

Note that if a prime number appears in the factorization of only one of the numbers, then the other exponent will be 0. The LCM and HCF of the numbers are given by

$$LCM(a, b) = p_1^{\max(e_1, f_1)} p_2^{\max(e_2, f_2)} \cdots p_r^{\max(e_r, f_r)}$$

$$HCF(a, b) = p_1^{\min(e_1, f_1)} p_2^{\min(e_2, f_2)} \cdots p_r^{\min(e_r, f_r)}$$

Since $\max(e_k, f_k) + \min(e_k, f_k) = e_k + f_k$ for $1 \leq k \leq r$, it follows that

$$LCM(a, b) \cdot HCF(a, b) = ab$$

For example, $LCM(144, 540) = 2^4 {\cdot} 3^3 {\cdot} 5^1 = 2160$ and $HCF(144, 540) = 2^2 {\cdot} 3^2 {\cdot} 5^0 = 36$. It can be verified that $2160 \times 36 = 2^6 \cdot 3^5 \cdot 5^1 = 144 \times 540$.

However, this technique for finding the HCF requires us to know the prime factorization in advance. Sometimes it is not easy to obtain this. For example, if you are asked to find the HCF of 629 and 259, it is not at all obvious that $629 = 37 \times 17$ and $259 = 37 \times 7$. Indeed, no efficient algorithm is known for finding the prime factorization of large numbers, and this is the principle behind the encryption techniques used in various internet applications.

There is, however, a very efficient algorithm to find the HCF of two numbers

9

without knowing their prime factorization. This is known as **Euclid's algorithm** and is one of the oldest algorithms in existence. Recall that if a and b are two positive integers, we can write $a = qb + r$ where $0 \leq r < b$. Essentially q is the quotient upon dividing a by b and r is the remainder.

Euclid's ingenuity was to observe that $HCF(a, b) = HCF(b, r)$. Indeed, if g is a common factor of a and b, it must also be a factor of $r = a - qb$. Thus any common factor of a and b must be a factor of r. Likewise, if h is a common factor of b and r, it must also be a factor of $a = qb + r$. Thus any common factor of b and r must also be a factor of a. Therefore, $HCF(a, b) = HCF(b, r)$.

But now we can repeat this procedure, writing $b = cr + s$ with $0 \leq s < r$ and deducing that $HCF(b, r) = HCF(r, s)$. The sequence $b, r, s, \ldots$ is a decreasing sequence of non-negative integers and must eventually hit 0. But $HCF(x, 0) = x$ for any positive integer x. We illustrate this with an example below.

$$
\begin{aligned}
HCF(629, 259) &= HCF(259, 111) = HCF(111, 37) \\
&= HCF(37, 0) = 37 \\
LCM(629, 259) &= \frac{629 \times 259}{HCF(629, 259)} = 4403
\end{aligned}
$$

1.3 Rational Numbers

As we have seen, the set of integers $\mathbb{Z}$ is not closed under division. Although 1 and 2 are both integers, $\frac{1}{2}$ is not an integer. However, we can build a new set that contains integers as well as their ratios. This set is called the set of **rational numbers**, and is usually denoted $\mathbb{Q}$.

A rational number is a number that can be expressed in the form $\frac{p}{q}$ where p and q are integers and $q \neq 0$. A rational number $\frac{p}{q}$ is said to be in standard form if p and q have no common factor other than 1 and $q > 0$. Every rational number can be reduced to a unique standard form. For example, the standard form of $\frac{6}{15}$ is $\frac{2}{5}$.

Let a and b be two rational numbers, with $a < b$. There exist infinitely many rational numbers between them. To see this, suppose there were only finitely many rational numbers between a and b, and let c be the smallest of these. (Note that there are sets like $\{0.1, 0.01, 0.001, \ldots\}$ that do not have a smallest element, but all such sets are infinite.) Since c is between a and b, so is the average of a and c, namely $\frac{a+c}{2}$, which is also a rational number.

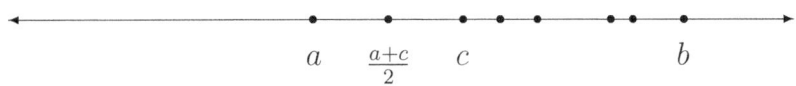

$$a \qquad \frac{a+c}{2} \qquad c \qquad\qquad b$$

But $\frac{a+c}{2}$ is smaller than c, so c is not the smallest rational number between a and b. The only way out of this situation is to concede that the number of rational numbers between a and b is infinite. ∎

Unlike integers, every non-zero rational number $\frac{p}{q}$ has a multiplicative inverse (also known as reciprocal), namely $\frac{q}{p}$. Rational numbers are essentially closed under division. The only exception is division by zero, which is not defined. The sum and product of the rational numbers $\frac{a}{b}$ and $\frac{c}{d}$ are the rational numbers $\frac{ad+bc}{bd}$ and $\frac{ac}{bd}$ respectively, though these may not necessarily be in the standard form. The concepts of additive identity, multiplicative identity and additive inverse that we encountered in the context of integers are equally applicable for rational numbers.

Every rational number has a terminating or recurring decimal expansion. For example,

$$\frac{11}{16} = 0.6875$$

$$\frac{11}{14} = 0.7857142857142... = 0.7\overline{857142}$$

Conversely, given a number with a terminating or recurring decimal expansion, we can recover its representation as a rational number.

Let $x = 0.6875$. Then $10^4 x = 6875$.

$$x = \frac{6875}{10^4} = \frac{625 \times 11}{625 \times 16} = \frac{11}{16}$$

Let $x = 0.7\overline{857142}$. Then $10x = 7.\overline{857142}$ and $10^7 x = 7857142.\overline{857142}$.

$$x = \frac{7857135}{10^7 - 10} = \frac{714285 \times 11}{714285 \times 14} = \frac{11}{14}$$

As seen above, the rational number $\frac{p}{q}$ written in standard form has a terminating expansion if and only if some multiple of q is a power of 10. It follows from the fundamental theorem of arithmetic that this will happen if and only if $q = 2^a 5^b$ where a and b are non-negative integers.

1.4   Decimals and Percentages

The subset of rational numbers where the denominator is a power of 10 is called the set of decimal numbers. The rules for decimal arithmetic are exactly the same as that of rational arithmetic. For example,

$$2.38 + 3.2 = \frac{238}{100} + \frac{32}{10} = \frac{558}{100} = 5.58.$$

$$2.38 \times 3.2 = \frac{238}{100} \times \frac{32}{10} = \frac{7616}{1000} = 7.616.$$

$$2.38 \div 3.2 = \frac{238}{100} \div \frac{32}{10} = \frac{2380}{3200} = 0.74375.$$

Since each rational number has infinitely many representations that are all equivalent to each other, it is not often apparent whether a rational number $\frac{p}{q}$ is greater than, equal to, or less than another rational number $\frac{r}{s}$. The most straightforward way to determine this is to test whether the difference $\frac{p}{q} - \frac{r}{s}$ is positive, zero or negative. However, to compare the magnitude of rational numbers with different denominators, it is often convenient to express them as percentages. Thus each rational number is rescaled to a denominator of 100 and the resulting numerator indicates the magnitude. Note that the numerator need not be an integer in this case.

For example, the rational numbers $\frac{3}{4}$ and $\frac{75}{100}$ correspond to the same number. The former is the standard form and the latter is the percentage form. We say that the fraction $\frac{3}{4}$ is equivalent to 75%. In general, the fraction $\frac{p}{q}$ is equivalent to $\frac{100p}{q}$ %.

Percentages obey the same rules as fractions. Taking $x\,\%$ of y is the same as multiplying y by the fraction $\frac{x}{100}$. When the value of a quantity increases from x by $y\%$, the result is obtained by adding to x, the number $y\%$ of x. When the value of the same quantity decreases by $y\%$, the result is obtained by subtracting from x, the number $y\%$ of x.

There is also a subtle distinction between the terms *increase by $x\%$* and *increase to $x\%$*. For example, when the sides of a square are doubled, we say that the area has increased *to* 400% and increased *by* 300%. When the sides are halved, we say that the area has decreased to 25% and decreased by 75%.

An interest rate of $R\%$ per annum on a principal of P essentially means that the value of the investment or loan is multiplied by $(1 + \frac{R}{100})$ at the end of the year. In simple interest, the base figure for computing interest is always the original principal. Thus the value of the deposit or loan at the end of N years is given by

$$A = P\left(1 + \frac{NR}{100}\right)$$

If interest is compounded annually, the accumulated amount including interest will be the base figure for computing next year's interest. Thus the value of the deposit or loan at the end of N years is given by

$$A = P\left(1 + \frac{R}{100}\right)^N$$

If the interest is compounded semi-annually, we replace 100 by 200 and N by $2N$ as the number of terms has doubled and the interest rate per term has halved.

15

For example, the value of a deposit or loan of 10,000 at the end of a period of two years at 6 % interest will be 11,200 for simple interest, 11,236 for interest compounded annually and 11,255.09 for interest compounded semi-annually.

1.5 Real Numbers

If the decimal expansion of a number is neither terminating nor recurring, it cannot be expressed as a rational number. Such numbers are called **irrational numbers**. For example, 0.101001000100001... is irrational. Rational and irrational numbers are collectively called **real numbers**. Every real number can be associated with a unique point on the number line.

We will now demonstrate that $\sqrt{2}$ is irrational. This is one of the oldest and most beautiful pieces of mathematical reasoning in existence, and was known to the ancient Greeks. It is an example of a technique known as **proof by contradiction**. We have already encountered this technique when we showed that there are infinitely many rational numbers between any pair of distinct rational numbers. The approach is as follows.

- We assume that the given statement is false.

- This assumption leads to an inescapable contradiction.

- We conclude that the statement must be true.

The great English mathematician G. H. Hardy, in his classic book *A Mathematician's Apology*, remarked that proof by contradiction is far superior to anything that happens in a game of chess, as "a chess player may offer the sacrifice of a pawn or even a piece, but a mathematician offers *the game*".

Suppose $\sqrt{2}$ is rational and can be written as $\frac{p}{q}$ in standard form. Thus p and q have no common factor other than 1.

$$\frac{p}{q} = \sqrt{2}$$

$$\frac{p^2}{q^2} = 2$$

Thus $p^2 = 2q^2$, so p^2 is an even number. Since the square of an odd number is an odd number, p must be an even number. Let $p = 2k$. Then $2q^2 = p^2 = (2k)^2 = 4k^2$, so that $q^2 = 2k^2$. Thus q^2 is an even number. So q must be an even number as well.

But $\frac{p}{q}$ was already in standard form, so p and q cannot both be even numbers, as they will then have the common factor 2. This contradiction shows that $\sqrt{2}$ is irrational. ∎

This argument can be easily modfied to show, for example, that $\sqrt{17}$ is irrational. Simply replace "even number" by "multiple of 17".

It may be worthwhile to point out that the following technique, which bears a superficial resemblance to proof by contradiction, has no place in mathematical reasoning and should be avoided at any cost.

- We assume that the given statement is true.

- As far as we can see, this assumption does not lead to any contradiction.

- We conclude that the statement must be true.

This is not proof by contradiction. This is proof by laziness. This is putting the cart before the horse. No marks for doing this.

Note that unlike rational numbers, irrational numbers are not closed under any of the arithmetic operations. Let $a = \sqrt{2}, b = 1 + \sqrt{2}$ and $c = 1 - \sqrt{2}$. All three numbers are irrational, but $b + c = 2, b - a = 1$ and $bc = -1$ are rational. Real numbers, on the other hand, obey the same closure properties as rational numbers.

Recall from our discussion on the laws of exponents that $\sqrt{a} = a^{\frac{1}{2}}$. It follows that $\sqrt{ab} = \sqrt{a} \times \sqrt{b}$ and

$$\sqrt{\frac{a}{b}} = \frac{\sqrt{a}}{\sqrt{b}}$$

Upon dividing one irrational number by another, it is often convenient to express the fraction in such a way that the denominator is rational. If the denominator contains an expression of the form $a + b\sqrt{c}$ where a, b and c are integers, we can multiply the numerator and denominator by $a - b\sqrt{c}$, so that the denominator becomes the integer $a^2 - b^2 c$. For example,

$$\frac{2}{9 - 3\sqrt{5}} = \frac{2}{9 - 3\sqrt{5}} \cdot \frac{9 + 3\sqrt{5}}{9 + 3\sqrt{5}} = \frac{18 + 6\sqrt{5}}{36} = \frac{3 + \sqrt{5}}{6}$$

This technique is called **rationalization of the denominator**.

1.6 Arithmetic Progressions

In a primary school in the German province of Brunswick in the 1780s, a frustrated mathematics teacher named J. G. Büttner had finally come up with a brilliant plan to keep his students quiet. He was particularly vexed by young Carl Friedrich, who put him at his wit's end.

"Class, find the sum of all the whole numbers from 1 to 100."

"Ligget se!" ("There it is!"), said Carl Friedrich a few seconds later and walked up to his teacher with a single number written on his slate. The number was 5050. The ten-year-old boy had independently discovered a formula for the sum of the first n positive integers. To nobody's surprise, he grew up to become the greatest mathematician of the modern era. His last name was Gauss.

An **arithmetic progression** or A.P. is a sequence of real numbers with the property that the difference between consecutive terms is a constant quantity, known as the **common difference**. The first term of an arithmetic progression is usually denoted a and the common difference is denoted d. Thus the second term is $a + d$, the third term is $a + 2d$, and in general, the nth term is $a + (n - 1)d$. The sequence $1, 2, 3, \ldots, 100$ is an arithmetic progression whose first term and common difference are both equal to 1.

Suppose a, b and c are three consecutive terms of an arithmetic progression. Since the difference between consecutive terms is constant, we have $b - a = c - b$. Thus $a + c = 2b$, and b is the average of a and c. In other words, any term in an arithmetic progression is the average of the previous term and the next term.

We will now derive, like Gauss did, a formula for the sum S of the first n terms of an arithmetic progression with first term a and common difference d. We will show that **the average of the terms of an arithmetic progression equals the average of the first and last terms**, from which the formula for the sum will follow. If n is odd, the average of all the terms is clearly the middle term, as it forms a 3-term AP of common difference d with its immediate neighbours, another 3-term AP of common difference $2d$ with terms two steps before and after, and so on. Thus

$$\frac{S}{n} = a + \left(\frac{n-1}{2}\right) d = \frac{2a + (n-1)d}{2}$$

$$S = \frac{n}{2}\left[2a + (n-1)d\right]$$

If n is even, then $n-1$ is odd, so to compute the sum we will add to a the sum of the first $n-1$ terms of the arithmetic progression with first term $a+d$ and common difference d. Thus

$$S = a + \left(\frac{n-1}{2}\right)\left[2(a+d) + (n-2)d\right] = \frac{n}{2}\left[2a + (n-1)d\right]$$

It turns out that the final answer does not depend on whether n is odd or even! Since the last term is $a + (n-1)d$, the average of the first n terms of an arithmetic progression equals the average of the first term and the last term. ∎

This formula has many interesting applications. For example, the sum of the first n positive integers is given by

$$S = \frac{n}{2}\left[2 + (n-1)\right] = \frac{n(n+1)}{2}$$

Putting $n = 100$, we get $S = 5050$, the number on the slate of young Carl Friedrich Gauss.

Similarly, the sum of the first n odd numbers is given by

$$S = \frac{n}{2}\Big[2 + 2(n-1)\Big] = n^2$$

One way to visualize this identity is to think of filling an $n \times n$ chessboard using L-shaped tiles, starting from the top right corner.

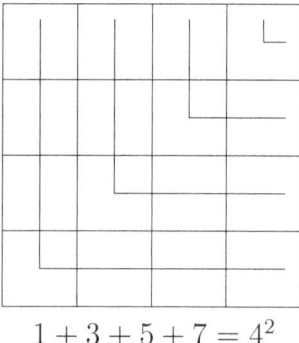

$$1 + 3 + 5 + 7 = 4^2$$

Thus any odd number can be written as the difference of two consecutive squares. For example, $17 = (1 + 3 + \ldots + 17) - (1 + 3 + \ldots + 15) = 9^2 - 8^2$. What this also means is that given an odd integer m, we can generate a right triangle with m as one of the sides by writing

$$
\begin{aligned}
m^2 &= (1 + 3 + \ldots + m^2) - (1 + 3 + \ldots + m^2 - 2) \\
&= \left(\frac{m^2 + 1}{2}\right)^2 - \left(\frac{m^2 - 1}{2}\right)^2
\end{aligned}
$$

We say that $\left(\frac{m^2-1}{2}, m, \frac{m^2+1}{2}\right)$ is a Pythagorean triplet for all odd integers m.

What about Pythagorean triplets for even integers? If m is an even integer, m^2 is a multiple of 4. Thus $m^2 = 4k = (2k-1) + (2k+1)$, so that

$$m^2 = (1 + 3 + \ldots + 2k + 1) - (1 + 3 + \ldots + 2k - 3) = (k+1)^2 - (k-1)^2$$

Thus we obtain the Pythagorean triplet $\left(\frac{m^2-4}{4}, m, \frac{m^2+4}{4}\right)$ for all even integers m.

Not all Pythagorean triplets are of this form. When does a triple of positive integers (a, b, c) form a Pythagorean triplet? Before we answer this, observe that whenever (a, b, c) forms a Pythagorean triplet, so does (da, db, dc) for any positive integer d. Thus it is sufficient to consider only *primitive* Pythagorean triplets (a, b, c), where $HCF(a, b, c) = 1$. The general form of a primitive Pythagorean triplet is $(m^2 - n^2, 2mn, m^2 + n^2)$ where m and n are positive integers with $m > n$ and $HCF(m, n) = 1$. Taking $m = 5$ and $n = 2$, we get the triplet $(21, 20, 29)$ which cannot be generated by either of the earlier formulas. The following table lists all Pythagorean triplets with largest entry less than 50.

23

m	n	$a = m^2 - n^2$	$b = 2mn$	$c = m^2 + n^2$
2	1	3	4	5
3	1	8	6	10
3	2	5	12	13
4	1	15	8	17
4	3	7	24	25
5	1	24	10	26
5	2	21	20	29
5	3	16	30	34
5	4	9	40	41
6	1	35	12	37

To illustrate another application, let us find the sum of all multiples of 7 between 100 and 300. The required set of numbers form an arithmetic progression with first term 105, common difference 7 and last term 294. Since $105 + 7(n - 1) = 294$, it follows that $n = 28$. Thus the sum of all multiples of 7 between 100 and 300 equals $14 \times (105 + 294) = 5586$.

Chapter 2

Algebra

2.1 Linear Equations

Among the oldest equations known to mankind is one that appears in *Rhind Papyrus*, an ancient Egyptian manuscript written around 1650 BC. It says: "A number added to one-seventh of itself yields 19."

In modern notation, this is written as

$$x + \frac{x}{7} = 19$$

Equivalently,

$$\frac{8x}{7} - 19 = 0$$

Any such equation of the form $ax + b = 0$, where a and b are real numbers and $a \neq 0$ is called a **linear equation in one variable** or a **simple equation**. The equation has exactly one solution, namely $x = -b/a$, that can be obtained by first subtracting b from both sides of the equation and then dividing both sides by a. In particular, the number with the property mentioned in *Rhind Papyrus* is

$$x = \frac{19 \times 7}{8} = 16.625.$$

Many equations that do not appear to be linear can be reduced to the above form. For example, if we are asked to find the age of a man who is three times as old as his son now, and was five times as old as the same son ten years ago, we can solve the equation $3x - 10 = 5(x - 10)$ where x is the present age of the son. Simplifying, we get $x = 20$ and the father's age to be 60 years.

You may know from physics that x degrees Celsius is the same as $\frac{9x}{5} + 32$ degrees Fahrenheit. At what temperature do the readings on the two scales coincide? The answer can be obtained by solving the equation

$$x = \frac{9x}{5} + 32$$

$$\frac{4x}{5} + 32 = 0$$

$$x = \frac{-32 \times 5}{4} = -40$$

What about linear equations in two variables x and y? We adopt the convention that a linear equation in two variables must be of the form $ax + by + c = 0$ where at least one of a and b is non-zero.

A linear equation in two variables has infinitely many solutions. To see this, consider the equation $ax + by + c = 0$ with $a \neq 0$. Observe that $(\frac{-c}{a}, 0)$ is a solution to the equation. Moreover, if (p, q) is a solution, so is $(p - b, q + a)$. Repeating this process, we can generate infinitely many solutions. All these solutions lie along a line when viewed as points on a plane.

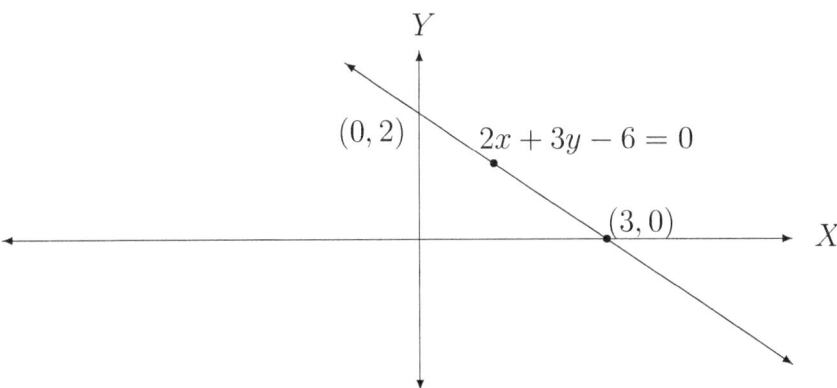

Since a simple equation can also be viewed as a special case of a linear equation in two variables, the graphs of the equations $x = a$ and $y = b$ are also straight lines. The former is a vertical line (parallel to the Y-axis) and the latter is a horizontal line (parallel to the X-axis).

2.2 Pair of Equations in Two Variables

We have seen that solutions of a linear equation in two variables correspond to points on a straight line. If we have two such equations $ax + by + c = 0$ and $px + qy + r = 0$, will there be any solution satisfying both equations? It turns out that there are three cases, each discussed in detail below.

If the lines are identical, then there will be infinitely many common solutions. For example, the equations $x + 2y - 6 = 0$ and $2x + 4y - 12 = 0$ are identical, as the second equation is just the first equation multiplied by 2 on both sides. Thus the corresponding lines are identical as well, and any of the infinitely many solutions for the first equation are solutions for the second equation and vice versa. This happens if and only if $\mathbf{aq - bp = 0}$ **and** $\mathbf{ar - cp = 0}$.

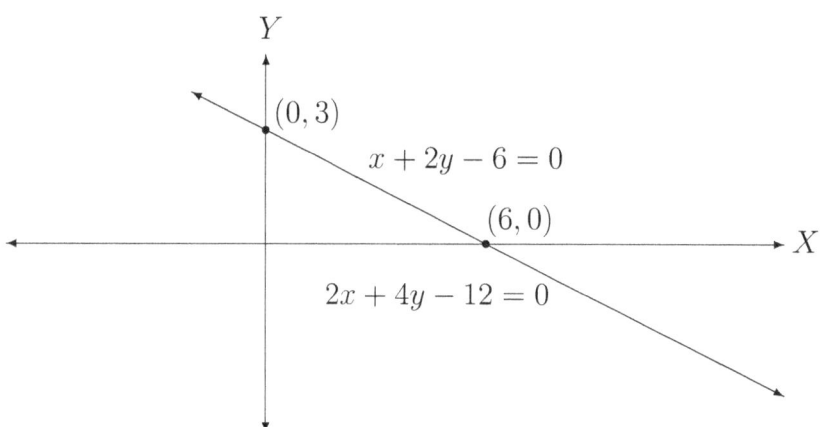

If the lines are parallel, then there will be no common solution. For example, the equations $x + 2y - 6 = 0$ and $2x + 4y - 8 = 0$ cannot have a common solution, as any solution to $x + 2y - 6 = 0$ will automatically satisfy $2x + 4y - 8 = 4$. Thus the corresponding lines will be parallel. This happens if and only if $\mathbf{aq - bp = 0}$ and $\mathbf{ar - cp \neq 0}$.

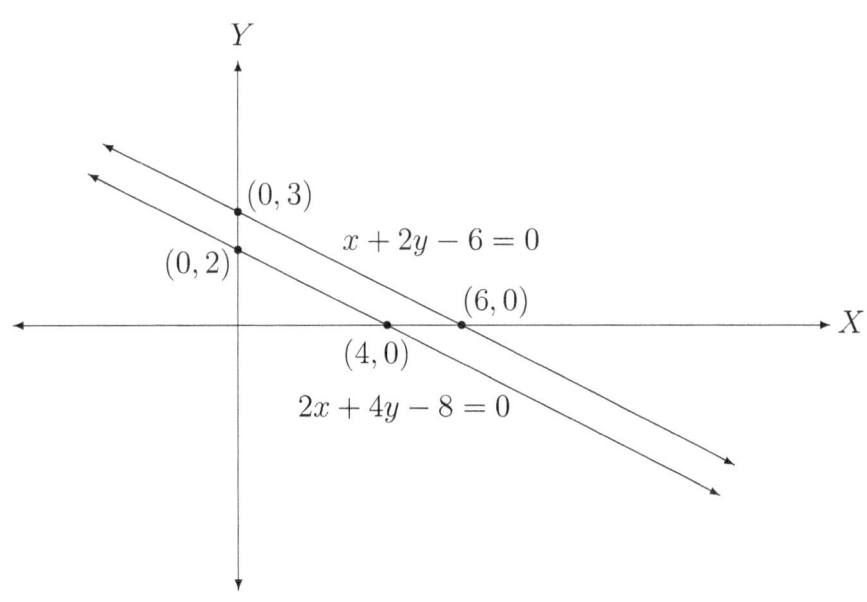

If the lines are neither identical nor parallel, they must meet at exactly one point, and there will be a unique common solution. For example, the lines corresponding to the equations $x + 2y - 6 = 0$ and $2x + y - 6 = 0$ meet at the point $(2, 2)$. Thus the only common solution to the two equations is $x = 2, y = 2$. This happens if and only if $\mathbf{aq - bp \neq 0}$.

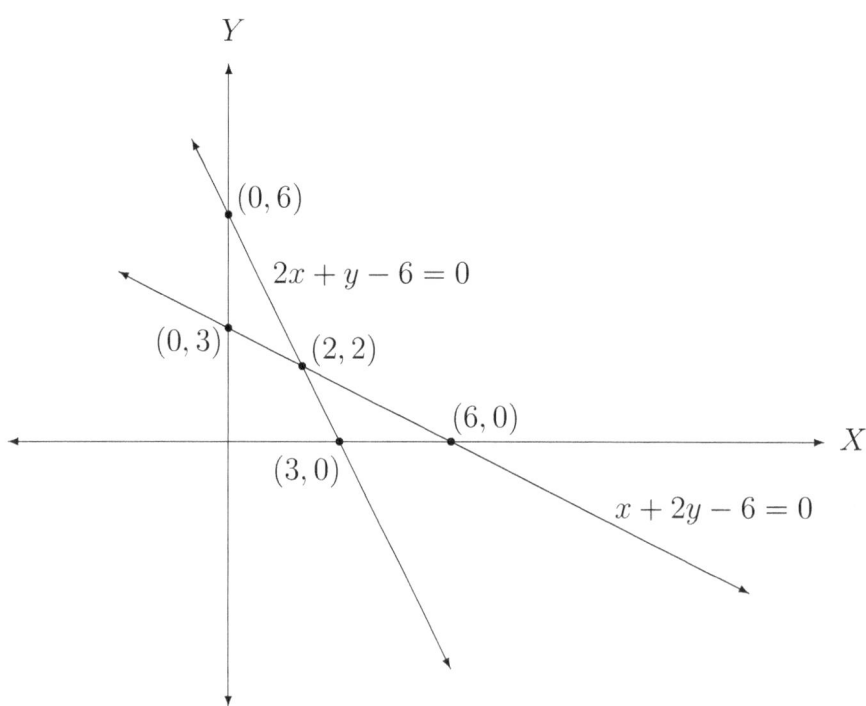

While it is possible to solve such systems graphically, it is often more efficient to do so algebraically. Let us consider an example. My mother has twice as many brothers as she has sisters, but her brother has the same number of brothers as he has sisters. How many brothers and sisters does my mother have?

Suppose my mother has x brothers and y sisters. Then her brother will have $x - 1$ brothers and $y + 1$ sisters. It is given that

$$x = 2y$$

$$x - 1 = y + 1$$

If we substitute the first equation in the second, we get $2y - 1 = y + 1$, i.e., $y = 2$. Now the first equation yields $x = 4$. Thus my mother has four brothers and two sisters.

Let us take another example. Can you find a two-digit number whose value increases by 27 when the digits are reversed, and whose sum of digits is 13?

Let the number be $10x + y$ where y is the digit in the unit's place and x is the digit in the tens' place. It is given that

$$10y + x = 10x + y + 27$$

Rearranging the terms and dividing throughout by 9, we get

$$y - x = 3$$

It is also given that

$$x + y = 13$$

Eliminating x by adding the two equations, we get $2y = 16$ and $y = 8$. Since the sum of digits is 13, we have $x = 5$. Thus the required number is 58.

Though the techniques presented in the above examples are essentially equivalent, some authors call the first one *the substitution method* and the second one *the elimination method*. Either of them can be used as the situation warrants.

2.3 Polynomials

A polynomial in x of degree n is an algebraic expression of the form

$$p(x) = a_n x^n + a_{n-1} x^{n-1} + \cdots + a_1 x + a_0$$

where $a_n, a_{n-1}, \ldots, a_0$ are real numbers and $a_n \neq 0$. In particular, real numbers are also polynomials. All non-zero real numbers have degree 0 as polynomials, whereas the degree of the zero polynomial is not defined. As in the case of real numbers, addition and multiplication of polynomials are commutative and associative, with the zero polynomial as additive identity and the constant polynomial 1 as multiplicative identity. Moreover, multiplication is distributive over addition. However, no polynomial with positive degree has a multiplicative inverse.

We say that the real number a is a **zero of the polynomial** $p(x)$ if $p(a) = 0$. In this case, we also say that $x = a$ is a **root of the equation** $p(x) = 0$. A polynomial of degree n has at most n zeroes.

If $p(x)$ and $g(x)$ are non-zero polynomials, then there exist unique polynomials $q(x)$ and $r(x)$ such that $p(x) = q(x)g(x) + r(x)$. Moreover, the degree of $r(x)$ is less than the degree of $g(x)$. We say that $q(x)$ is the quotient upon dividing $p(x)$ and $r(x)$ is the remainder.

For example, let $p(x) = x^4 + x + 1$ and $g(x) = x^2 + 1$. Then we have

$$p(x) = x^4 + x^2 - x^2 + x + 1 = x^2 g(x) - x^2 + x + 1 = (x^2 - 1)g(x) + (x + 2)$$

Thus $p(x) = q(x)g(x) + r(x)$ where $q(x) = x^2 - 1$ and $r(x) = x + 2$.

If $g(x) = x - a$, the degree of $g(x)$ equals 1, so the degree of $r(x)$ must be 0. In other words, the remainder is a constant real number, independent of the value of x. To determine this value, we put $x = a$ in the equation $p(x) = q(x)(x - a) + r(x)$ and obtain $r(x) = p(a)$. Thus **the remainder obtained on dividing $\mathbf{p(x)}$ by $\mathbf{x - a}$ equals $\mathbf{p(a)}$.**

If $p(a) = 0$, i.e., the real number a is a zero of $p(x)$, then the remainder obtained on dividing $p(x)$ by $x - a$ equals 0. Thus $\mathbf{p(a) = 0}$ **if and only if $\mathbf{x - a}$ is a factor of $\mathbf{p(x)}$.**

For example, it can be verified that $p(x) = 2x^3 - 16x^2 + 34x - 20$ satisfies $p(1) = p(2) = p(5) = 0$. Thus $p(x) = c(x - 1)(x - 2)(x - 5)$. By comparing the coefficient of x^3 on both sides, we get $c = 2$. It follows that

$$2x^3 - 16x^2 + 34x - 20 = 2(x - 1)(x - 2)(x - 5)$$

To find the zeroes of a polynomial, it is enough to solve the equations that

34

represent the coefficients in terms of the roots. For example,

$$(\mathbf{x} - \mathbf{a})(\mathbf{x} - \mathbf{b}) = \mathbf{x^2} - (\mathbf{a} + \mathbf{b})\mathbf{x} + \mathbf{ab}$$

$$(\mathbf{x} - \mathbf{a})(\mathbf{x} - \mathbf{b})(\mathbf{x} - \mathbf{c}) = \mathbf{x^3} - (\mathbf{a} + \mathbf{b} + \mathbf{c})\mathbf{x^2} + (\mathbf{ab} + \mathbf{bc} + \mathbf{ac})\mathbf{x} - \mathbf{abc}$$

However, there is no general formula to determine the zeroes of a polynomial of degree five or higher. The discovery that no such formula can possibly exist is one of the greatest achievements of modern mathematics. Unfortunately, the two young mathematicians who made the most significant contributions to this discovery, Evariste Galois and Niels Abel, both died in their twenties.

We now use the algebraic identities that we are already familiar with to derive some new ones. Recall that the identity $(a + b)^2 = a^2 + 2ab + b^2$ is about computing the square of a binomial. We can use it to compute the square of a trinomial.

$$(x + y + z)^2 = (x + y)^2 + 2z(x + y) + z^2$$

$$(\mathbf{x} + \mathbf{y} + \mathbf{z})^2 = \mathbf{x^2} + \mathbf{y^2} + \mathbf{z^2} + \mathbf{2xy} + \mathbf{2yz} + \mathbf{2xz}$$

We can also compute the cube of a binomial.

$$(x + y)^3 = (x + y)(x^2 + 2xy + y^2) = x^3 + 3x^2y + 3xy^2 + y^3$$

$$(\mathbf{x} + \mathbf{y})^3 = \mathbf{x^3} + \mathbf{y^3} + \mathbf{3xy}(\mathbf{x} + \mathbf{y})$$

$$\mathbf{x^3} + \mathbf{y^3} = (\mathbf{x} + \mathbf{y})(\mathbf{x^2} - \mathbf{xy} + \mathbf{y^2})$$

35

$$(\mathbf{x} - \mathbf{y})^3 = \mathbf{x}^3 - \mathbf{y}^3 - \mathbf{3xy}(\mathbf{x} - \mathbf{y})$$

$$\mathbf{x}^3 - \mathbf{y}^3 = (\mathbf{x} - \mathbf{y})(\mathbf{x}^2 + \mathbf{xy} + \mathbf{y}^2)$$

These identities are useful in factorising numbers. For example, it is clear that 1000027 is not a prime number, since it can be written as $100^3 + 3^3$ and must therefore be a multiple of 103. No such identity holds for squares. Indeed, 10009 is a prime number.

Computing the cube of a trinomial yields another identity.

$$(x + y + z)^3 = (x + y)^3 + z^3 + 3z(x + y)(x + y + z)$$

$$(x + y + z)^3 = x^3 + y^3 + z^3 + 3xy(x + y) + 3z(x + y)(x + y + z)$$

$$(x + y + z)^3 = x^3 + y^3 + z^3 - 3xyz + 3xy(x + y + z) + 3z(x + y)(x + y + z)$$

$$x^3 + y^3 + z^3 - 3xyz = (x + y + z)\left[(x + y + z)^2 - 3xy - 3yz - 3xz)\right]$$

$$\mathbf{x}^3 + \mathbf{y}^3 + \mathbf{z}^3 - \mathbf{3xyz} = (\mathbf{x} + \mathbf{y} + \mathbf{z})(\mathbf{x}^2 + \mathbf{y}^2 + \mathbf{z}^2 - \mathbf{xy} - \mathbf{yz} - \mathbf{xz})$$

2.4   Quadratic Equations

An equation of the form $ax^2 + bx + c = 0$ with $a \neq 0$ is called a quadratic equation. By dividing throughout by a, we see that the equation can be reduced to the form $x^2 + px + q = 0$ where $p = \frac{b}{a}$ and $q = \frac{c}{a}$. This latter form, with the leading coefficient equal to 1, is called the *monic form*.

We know that a polynomial of degree n has at most n zeroes. Thus a quadratic equation can have zero, one or two roots. The equation $x^2 - 7x + 12 = 0$ has two roots, namely 3 and 4. The equation $x^2 - 4x + 4 = 0$ has only one root, namely 2. The equation $x^2 + 1 = 0$ has no roots.

To determine the roots of a quadratic equation $x^2 + px + q = 0$ in monic form, it suffices to find real numbers u and v, if they exist, with the property that $x^2 + px + q = (x - u)(x - v)$. Comparing coefficients, we find that $u + v = -p$ and $uv = q$. Thus the essential problem of solving a quadratic equation reduces to finding two numbers with a given sum and a given product.

Can we do this? If I tell you that the sum of two numbers u and v is 31 and their product is 228, can you determine the numbers?

We proceed as follows. Since $u + v = 31$, we have

$$(u + v)^2 = u^2 + 2uv + v^2 = 31^2 = 961$$

37

$$(u - v)^2 = u^2 - 2uv + v^2 = (u + v)^2 - 4uv = 961 - 4 \times 228 = 49$$

$$u - v = \pm 7$$

Now that we know the sum and the difference, it is easy to find the numbers. If $u - v = 7$, we get $u = 19$ and $v = 12$. If $u - v = -7$, we get $u = 12$ and $v = 19$. Since addition and multiplication are commutative, we should not be able to say which of u and v is 12 and which is 19, and indeed we cannot.

In general, we are given $u + v = -p = -\frac{b}{a}$ and $uv = q = \frac{c}{a}$. Thus

$$(u - v)^2 = (u + v)^2 - 4uv = \frac{b^2 - 4ac}{a^2}$$

$$\frac{u - v}{2} = \pm \frac{\sqrt{b^2 - 4ac}}{2a}$$

Since $\frac{u+v}{2} = -\frac{b}{2a}$, the roots of the quadratic equation $ax^2 + bx + c = 0$ are

$$\mathbf{\frac{-b \pm \sqrt{b^2 - 4ac}}{2a}}$$

The quantity $b^2 - 4ac$ is called the **discriminant** of the quadratic equation. **If the discriminant is negative, the quadratic equation has no roots. If the discriminant is zero, the equation has only one root. If the discriminant is positive, the equation has two distinct roots.**

The following problem is from the Indian mathematician Bhaskaracharya's treatise *Lilavati*, written around 1150 AD. "O learned one! Find the number which when added to nine times its square root gives 1240."

Let this number be x^2. Then we have $x^2 + 9x - 1240 = 0$. The roots of this equation are

$$x = \frac{-9 \pm \sqrt{81 + 4960}}{2} = 31 \text{ or } -40$$

Are both solutions admissible? Not really. 961 added to 9 times 31 does give 1240. One could argue that 1600 added to 9 times -40 also gives 1240, but then -40 is not the square root of 1600. It is a solution to the quadratic equation $x^2 - 1600 = 0$, but the term *square root* is reserved for the non-negative candidate. The number which when *diminished* by nine times its square root gives 1240 is indeed 1600, but the only answer to the given question is 961.

An alternative technique for solving a quadratic equation like $x^2 + 9x = 1240$ is called **completing the square**. We observe that adding $\left(\frac{9}{2}\right)^2$ to both sides of the equation gives

$$\left(x + \frac{9}{2}\right)^2 = 1240 + \frac{81}{4} = \left(\frac{71}{2}\right)^2$$

Thus $x + \frac{9}{2} = \pm\frac{71}{2}$ and $x = 31$ or -40.

2.5 Playing With Numbers

And now for something completely different. In the following sum, each letter stands for a different digit. Moreover, **FOUR** is a multiple of 4, **FIVE** is a multiple of 5 and **NINE** is a multiple of 3. Determine what digit each letter denotes.

$$F\,O\,U\,R\,+$$

$$F\,I\,V\,E$$

$$------------$$

$$N\,I\,N\,E$$

The solution is given on the next page for your convenience.

$$\textbf{F O U R} +$$

$$\textbf{F I V E}$$

$$\text{------------}$$

$$\textbf{N I N E}$$

It is clear that $\textbf{R} = 0$. Since each letter stands for a different digit and $\textbf{FIVE}$ is a multiple of 5, we must have $\textbf{E} = 5$.

Since $\textbf{O} \neq 0$, the addition of $\textbf{U}$ and $\textbf{V}$ must result in a carry of 1, and this carry when added to $\textbf{O}$ and $\textbf{I}$ should yield $10 + \textbf{I}$. This new carry gives us the equation $2\textbf{F} + 1 = \textbf{N}$. Thus $\textbf{O} = 9$ and $\textbf{N}$ must be odd.

We also have $\textbf{U} + \textbf{V} = 10 + \textbf{N}$. Since 5 and 9 are already taken, $\textbf{N} = 3$ or $\textbf{N} = 7$. But $\textbf{N} = 7$ would require either $\textbf{U} = 9$ or $\textbf{V} = 9$, both of which are impossible. Thus $\textbf{N} = 3$ and $\textbf{F} = 1$.

Now $\textbf{U} + \textbf{V} = 13$. Since 5 and 9 are taken, $\textbf{U}$ and $\textbf{V}$ must be 6 and 7 in some order. Since $\textbf{FOUR}$ is a multiple of 4, we have $\textbf{U} = 6$ and $\textbf{V} = 7$.

The only possibilities remaining for $\textbf{I}$ are $2, 4$ and 8. Since $\textbf{NINE}$ is a multiple of 3, the sum of its digits must be a multiple of 3. Thus $\textbf{I} = 4$. This completes the solution.

1 9 6 0 +

1 4 7 5

----

3 4 3 5

Chapter 3

Statistics and Probability

3.1 Measures of Central Tendency

Albert Einstein once said, "Everything should be made as simple as possible, but not simpler." Rather than deal with a large collection of numbers, it is often convenient to deal with a single number that captures the spirit of the collection. Such a choice should be *representative* of the numbers. Choosing the largest or smallest number in the collection, for example, is not a good idea. Three such choices have turned out to be quite popular, and these are the **mean**, the **median** and the **mode**.

As an example, let us take the first 25 terms of the sequence of gaps between consecutive prime numbers.

$$1, 2, 2, 4, 2, 4, 2, 4, 6, 2, 6, 4, 2, 4, 6, 6, 2, 6, 4, 2, 6, 4, 6, 8, 4$$

In order to determine the median, we must sort the sequence in increasing order of terms. The numbers 1 and 8 occur only once, while 2, 4 and 6 occur 8, 8 and 7 times respectively. Since the number of terms is odd, there is a unique middle term which equals the median. (If the number of terms were even, the median would be the average of the two middle terms.) In this case, the median is the thirteenth term of the *sorted* sequence, namely 4.

The mean is just the average of all the terms, and there is no need of sorting. In this case, the average equals

$$\frac{1 \times 1 + 2 \times 8 + 4 \times 8 + 6 \times 7 + 8 \times 1}{25} = 3.96$$

Of course, someone who knows how this sequence is constructed can easily compute the mean to be $\frac{101-2}{25} = 3.96$, but in practice such information is rarely available.

There are two values, namely 4 and 6, that occur with the highest frequency. Thus the data is *multimodal*, with modes 4 and 6.

All this easy. The more interesting question is, "Which of these three choices is truly representative of the data?"

The answer depends a lot on context. The mode is generally useful in a voting scenario. After all, if the data consists of choices from an electronic voting machine recorded against the serial numbers of candidates, then the mode does correspond to the winning candidate. It is in this "democratic" sense that mode is representative.

The mean is like the centre of mass in physics. It is the point where the weight of the entire data can be assumed to be concentrated. Data points with large magnitude tend to pull the mean towards themselves, much like the head of a hammer. If you want the influence of a data point to be proportional to its magnitude, go with the mean.

The median is what we call a more "robust" measure than the mean, as it is less sensitive to extreme values that occur infrequently, called *outliers*. Consider a small town of fifty households, where twenty households have a monthly income of 500 each, seventeen have an income of 3000 each, twelve have an income of 10,000 each and one household has a monthly income of 200,000. The mode is 500 and the mean is 7620, but the median of 3000 appears to be a better representative overall.

For our example of prime gaps, I have a preference for the mean. This is mainly because it is possible to say something meaningful about the mean in this case. One of the consequences of a deep theorem in number theory called the **prime number theorem** is that the mean of the first N terms of the sequence of gaps of prime numbers is approximately

$$1 + \frac{1}{2} + \frac{1}{3} + \cdots + \frac{1}{N}$$

For $N = 25$, the above approximation yields 3.82, or a relative error of less than 4%. Not bad at all.

However, nothing is known about the median or mode of the sequence of prime gaps. Indeed, we do not know of a single number in the sequence that is guaranteed to occur infinitely often. One of the most famous open problems in number theory is the **twin prime conjecture**, which states that 2 occurs infinitely often as a prime gap. An exciting development occurred in 2013 when a mathematician named Yitang Zhang showed that some even number between 2 and 70 million must occur infinitely often in the sequence of prime gaps. Subsequently, an army of mathematicians collaborated over the internet and brought down the upper bound from 70 million to 246, which is where things stand as of now.

3.2 Mean, Median and Mode of Grouped Data

When data is grouped into class intervals, the mean, median and mode can only be estimated. This is essentially done by assuming that the data is uniformly distributed within each class interval.

The computation of the mean is straightforward. Since we do not know any better, we assume that all the values in a class interval coincide with the midpoint of the class interval called the **class mark**. We have now reduced the grouped data problem to an ungrouped data problem.

For example, consider the following grouped data.

Class Interval	Class Mark	Frequency
$10-20$	15	3
$20-30$	25	1
$30-40$	35	4
$40-50$	45	2

The mean is given by

$$\frac{15 \times 3 + 25 \times 1 + 35 \times 4 + 45 \times 2}{10} = 30$$

To compute the median of the above data, we determine the leftmost class interval where the cumulative frequency exceeds half the total frequency. In this case, half the total frequency is 5 and the class interval with this property is $30-40$. The cumulative frequency at its left endpoint is 4 and at its right endpoint is 8. The median m is given by the solution to the simple equation

$$\frac{m - 30}{40 - m} = \frac{5 - 4}{8 - 5}$$

Solving the equation, we obtain the median to be 32.5. As expected, the median lies in the chosen class interval.

Since there is a unique class interval with highest frequency, called the **modal class**, we can compute the mode of the above data as well. Note that the modal class is the interval $30 - 40$ with frequency $f_1 = 4$. The frequency of the class preceding the modal class is $f_0 = 1$ and the frequency of the class succeeding the modal class is $f_2 = 2$. (We take $f_0 = 0$ if there is no preceding class and $f_2 = 0$ if there is no succeeding class.) The mode M is given by the solution to the simple equation

$$\frac{M - 30}{40 - M} = \frac{f_1 - f_0}{f_1 - f_2}$$

Thus the mode equals 36 and belongs to the modal class, as it should.

3.3 Probability of Events

Making decisions in the face of uncertainty is what growing up is really about. We all have intuitive mechanisms that help us with this, but in complex situations it is useful to have mathematics on our side. The branch of mathematics that deals with these topics is called **probability theory**.

The basic framework of probability theory consists of one or more trials, and the set of outcomes associated with each trial. Each individual outcome of the trial is called an **elementary event**. An event, in general, consists of zero or more elementary events. **All elementary events are assumed to be equally likely**. If an event **E** consists of M elementary events, and there are N possible outcomes in all, we say that the event **E** occurs with probability $\frac{M}{N}$.

For example, if we roll a die, there are six possible outcomes. The event that the number that appears on top is greater than 4 consists of two elementary events, namely the event that 5 appears, and the event that 6 appears. Thus the probability of this event equals $\frac{2}{6} = \frac{1}{3}$.

If we toss two fair coins, and consider the probability that exactly one of them shows head, there are two favourable outcomes, namely HT and TH out of the four possible outcomes, namely HH, HT, TH and TT. Thus the probability of this event equals $\frac{2}{4} = \frac{1}{2}$.

A common but incorrect line of reasoning for the above question is as follows.

The number of heads can be zero, one or two. Only one of these corresponds to a favourable event. Thus the probability is $\frac{1}{3}$. Even the 18th century French mathematician and physicist Jean d'Alembert fell into this trap before the axioms of probability were put on a sound footing by his student Pierre-Simon Laplace in 1814. What exactly is the mistake here?

Note that the three events, namely none of the coins showing head, exactly one of the coins showing head, and both coins showing head are not equally likely. The first and last are elementary events with probability $\frac{1}{4}$, while the second consists of two elementary events and has probability $\frac{1}{2}$ as we have just seen. In this case as well as in general, **the sum of probabilities of all elementary events equals 1**.

If **E** is an event, the **complementary event**, denoted $\overline{\mathbf{E}}$, consists of the set of outcomes where **E** does not occur. Since exactly one of **E** and $\overline{\mathbf{E}}$ must happen, and they cannot happen simultaneously, it follows that all the elementary events must belong to either **E** or $\overline{\mathbf{E}}$. Thus $P(\mathbf{E}) + P(\overline{\mathbf{E}}) = 1$. For example, the probability that the number appearing on top when a die is rolled is less than or equal to 4 is given by $1 - \frac{1}{3} = \frac{2}{3}$.

If an event consists of all possible outcomes, then it is certain to occur and has probability 1. Such an event is called a **certain event**. For example, the number appearing on top when is a die is rolled will certainly be less than 7, and this event has probability 1.

Similarly, if an event contains none of the possible outcomes, it will never occur and has probability 0. Such an event is called an **impossible event**. For example, the number on top when a die is rolled can never be greater than 6, and this event has probability 0.

Let us now consider another example. Suppose a box contains five red balls and five green balls. Two balls are drawn from the box, one after the other. What is the probability that the balls have the same colour?

After the first ball is drawn, there are only four more balls remaining in the box that are of the same colour as the first ball, whereas there are five balls of the other colour. Thus the probability is $\frac{4}{9}$.

If the first ball is put back in the box after it is drawn, then there are five balls of each colour in the box, and the probability of the second ball having the same colour as the first ball is $\frac{5}{10} = \frac{1}{2}$. In general, we say that the probabilities depend on whether sampling is done with or without replacement.

www.ingramcontent.com/pod-product-compliance
Lightning Source LLC
Chambersburg PA
CBHW081259180526
45170CB00007B/2495